The Gradual Implosion and Domino Effect in Florida

Joan M. Quitian Ramos Ramos

1. Authoritarian Education Systems: Ah, Florida, where the

educational system is a brilliant exercise in manufacturing a generation of docile conformists. It's a place where the quest for 'safety and order' miraculously translates into a populace stunningly ill-equipped for critical thinking. It's quite the achievement, really, molding minds that might struggle to navigate the complexities of a SpongeBob SquarePants episode, let alone

the global trade
network.

2. Psychological
Trauma and
Dissociation: In an
impressive twist,
students are
turning to astral

projection and
dissociative
disorders, not for
amusement, but
as survival
strategies against
an educational
regime that would
make Orwell

proud. This bouquet of psychological distress, nurtured by Florida's own brand of environmental stress, is brilliantly

preparing
individuals who
are about as ready
to tackle global
economic
challenges as a
cat is to solve
quantum physics.

3. Deterministic Viewpoints and Reductionism: The deterministic approach in Florida's policymaking is like watching someone use a

hammer for every problem – whether it's a nail or not. It's an adorable oversimplification, reducing the rich tapestry of human experience to mere chemical

reactions. It's as if
they believe
societal progress
is just another
item on a fast-food
menu – quick,
cheap, and
ultimately
unsatisfying.

Economic Ripple Effects and Global Trade Impacts

- Economic Instability: The crumbling edifice of Florida's

educational and societal structures is not just a local melodrama. It's a threat to its economic vitality, particularly amusing when you consider its self-imagined role as a linchpin in global trading routes.

- Global Trade Disruptions: As Florida grapples with its self-inflicted absurdities, it's unwittingly sabotaging its capacity to play in the big leagues of global trade. It's like watching a gambler betting his house on a

game he doesn't
know how to play.

**Perception and
Awareness**

- For the Unaware:
For those
engrossed in the

mundane cycle of
daily needs and
transient
pleasures, these
issues are as
distant as a galaxy
far, far away. Yet,
the societal
collapse,

masterfully
initiated by a
flawed
educational
system, will
eventually make
its grand entrance
into their lives,
like an uninvited

but inevitable
guest.

- For the Informed:
For the
enlightened few,
Florida's plight is a
sobering reminder
of how

interconnected our societal webs are. It's a showcase of how a police state's romance with education can lead to a psychological and

economic
hangover, the kind
that lingers.

Historical Context: Eugenics in Latin America (1920-1930)

- Learning from
History: The
eugenics
practices in early
20th century Latin
America, ethically
murky as they are,
offer a quaint
mirror to Florida's

current follies. It's
like looking back
at your awkward
teenage years,
except with more
at stake.

Addressing the Current Crisis in Florida

-Protocols for Assistance: Addressing Florida's woes involves a multi-

faceted approach
that might just be
too sensible for
the current
regime. It includes
psychological
support, a radical
idea of reform in
education and law

enforcement, and
– wait for it –
actually involving
the community.

**Broader
Implications**

**Global Impact: Florida's internal chaos isn't just a local soap opera; it's a global concern, like a small tear in a priceless painting

that risks ruining
the entire artwork.

In Conclusion

In essence,
addressing the
situation in Florida
is like trying to

clean up after a wild party that nobody admits to attending. It requires a strategy that's grounded in reality, not just the fantasies of those in power.

It's about time to wake up and smell the impending disaster – or continue to play the violin while Rome burns.

Personal Reflection

As a long-term resident with thwarted political aspirations, who has witnessed the tragicomic

downfall of a once-promising state, I've come to realize that expecting rationality in this context is like expecting a cat to bark. The current

state of affairs in
Florida is not just
a lost cause; it's a
cautionary tale of
what happens
when ignorance is
at the helm. As for
the architects of
this disaster, one

does wonder if
they yearn for the
blissful ignorance
of their own
prenatal state.

Joan Manuel Quitian Ramos.